This is Life
Conversations with God
Prelude to the Soul Camera

Larissa A Watson

This is Life

The big questions answered in
Conversations with God

Prelude to the Soul Camera

By

Larissa A Watson

Written in 2017 in Spain

Published in 2018 in Italian, Spanish, French, German and English

By

Larissa A Watson

Independent ISBN Code 97 81916421127

Introduction

This book has been written after a very challenging and painful time in my life during which my whole sense of self and reality fell apart.

During this time I struggled with my sanity, my faith in God, religion, man and beast. I have always believed in the existence of a higher power and it was at this time that I leaned heavily on this faith and in doing so my faith became a close and personal relationship with God, spirit, myself.

As a product of this close relationship and these in-depth conversations I have been guided to write this book as a way to help my fellow humans at this time of change.

The book was not intended as an instruction manual or a type of bible but as a way to help decipher some of the questions and themes that we might have of god if we were to sit down and try to explain how the deeper mysteries of the universe work and how we can help ourselves as humans move forward from this point in history.

I have divided the book into sections or possible questions to frame the information which I have been shown and instructed in. And I present it in a simple concise way. Many of these conversations and illuminations came as an organic, free flowing sequence of dreams, teachings and auditory instruction over the course of my life and more intensively over the past couple of years.

I was instructed to write this text although the thought terrified me. I literally took an anxiety attack at the prospect of doing this but I was assured that I would have help and assistance every step of the way to ensure only the highest truth would be written and that it would be done in a way which would have the highest level of support.

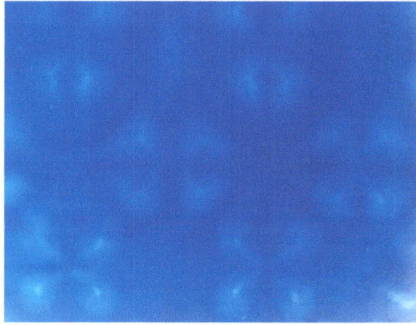

Contents

Is there a god?

Yes there is a god, but perhaps not in the way you might think. God is everywhere, is everything and everything is made of god. God is mind, is man and thought and love and spirit. If you accept this then let me explain;

I would like you to think for a moment about the idea of humanity as a being in the universe.

We are all made of matter, and the universe is matter and although we all are of the same we are all different. Our thoughts, experiences and environment shape us, as does the power of our emotions.

If you look at this proposition and understand that we are all made of the same source then we are all part of the all.

We are energy and matter. We are basically carbon atoms infused with the light and the energy of life and love.

We are not just carbon though, we are human and divine beings made by a higher power.

If we think in scientific terms we are chemical elements. However we exist in our own beings.

We think, therefore we are!

If we understand at a basic level that we are these basic elements that have somehow exploded into life as beautiful humans, walking, breathing thinking and doing beings then we also understand that we are more than atoms, particles, elements and stardust.

If we understand this then we must then know that we cannot be this amazing complex and extraordinary human being unless something happened to change chemistry into living matter.

If we know that all matter is stationary or potential until some energy, force or power gets involved to instigate a movement or reaction then we understand that we are therefore products of something greater than mere chemistry.

If we understand that, then we understand that something happened that was above and beyond our power to comprehend that put us here.

We are not mere strands of DNA; we are living, breathing units of perfect excellence,

We as an individual can act and behave as an entire universe in that we are more than our chemical makeup – we are more than just elements in the periodic table – we are imbued with something that is more, something that cannot be just chemistry. We have each been gifted with a divine spark of soul, light and life force and we are all part of the great creation. We are all part of our creator. We have been gifted with a part of the creation that made us.

If you consider that divine spirit is one living, breathing, moving, conscious organism and that us humans have all been made possible through having part of this spirit spark or seed in our being.

This therefore makes us part of the divine spirit source of God and creation.

I am of a creator background. I read and am very interested in physics, philosophy, mathematics, genetics, ecology and love life's pain and pleasures. I do not stand as an expert in these fields however I have done my work, I listen and converse with my gut, my guides, my higher self, power and am open to help from divine guidance. I give you this knowledge not from myself but because I was instructed to help others find answers in today's very conflicted and confusing world.

When I write these words I write with guidance and instruction both from a higher place and centred deep within me, from my own place. I have been tested and shown where my gaps in knowledge and understanding are and to strengthen my humanity as part of this journey. Also to assist me in knowing myself and knowing that everything is possible with self belief and trust in having the opportunity to see this.

We are magnificent beings who have been placed on this earth at this crucial time in history to make a difference and make a change possible.

The change required of us by our higher force or power is to become the best version of ourselves stunningly wondrous beings that we were destined to be so that we can bring to birth a new era and time for humanity and evolution for human kind. We are called upon to be the best most beautiful versions of ourselves possible.

There are a few fundamental truths and instructions that hold true for all of us.

These are;

We must know ourselves first before anything else.

We must learn to truly love who we are from a cellular level up to how we present in the world.

We are the creators of our thoughts, lives and environments and existence and we have the power to change this at will.

I would say at this stage that I know we are not the only life form in the universe and not the only life form on our planet. We are here by means of higher spirit and as such we are being watched over and guided on our daily lives to help us move forward.

1. Why am I here?

There have been many theories on this subject but I can only speak with which I have been taught and tutored.

There are spectrums of truth and I have been shown these on different levels. I have lived through them and experienced the energy and frequencies that each have been shown to me and from my experience I explain.

If you consider the universe as an entire spectrum ranging from the most animalistic, grotesque and dark elements of humanity or universal cosmic beings through to a pure white godlike nothingness where only perfection and divine thought exists then know that we are all somewhere on this spectrum. Our job is to create balance in the universe. Our role is to play out and explore all the facets and nuances of the elements of creation throughout our lives. We were all born with the power for greatness and godlike excellence and also with the potential for destruction and depravity.

It is through our choices and thoughts that we create our lives because in being part of the divine spark we were made in the image and likeness of God our creator who breathed life into our carbon potential. It is what we do with our thoughts and choices that we determine the quality of our reality and what we make of our lives.

Our thoughts and actions create our existence, which are guided by our beliefs. Through the help of our guides, the guides we were born into this world with and those which are sent to help us with specific tasks we make our reality.

If you consider that when we are born into this existence having this incredible power to create anything we want for our creation then it makes sense that we are not allowed to operate on our own entirely. We are accompanied by spirit guides, guardians and a massive support system to ensure that what we create does a minimal amount of damage to the larger scheme of creation and ourselves. If you consider it like a business or production with a CEO, a communications manager, a producer and director then the entirety of our lives is like a story with which we work to produce our very own masterpiece under expert tuition.

This plan for our lives is blank for us to create our individual realities and we are tested and proved to determine how much is true and built on solid ground and how much is illusion and built on sand.

There is the pure light side, that which is always light and which is always pure and good. This frequency is very high and to the mere human being , even to those whom are pure and true in spirit it is an extremely difficult place to operate from as the frequency can feel extremely acidic, nauseating and as though you are dissolving into an acid nothingness. There is no emotion or feeling, there is only energy.

When I first became aware of frequency and had elevated myself quite high on the spectrum I kept blacking out at this level and at certain points I would begin to feel as though I would dissolve into nothingness. Although I had always veered toward a righteous and pure place I found this to be a very uncomfortable place to be in.

The darker side of the spectrum is that place in which we discover our darker and basic nature. Depending on who you are this can vary. I did not really understand it until I was shown it up close. This experience and journey was very difficult for me. I have always lived on the sunny side of the street. In this place I could not find my sunny spot for a very long time. It all started when I got into a personal intimate relationship with someone who was very comfortable with their dark side and quite extreme, as I was in a very open and accepting stage in my journey I became privy to this in the most personal way. It was as though his darkest places became my daylight and I lived his nightmares. It felt like a mind transfer and I spent months and months dealing with the download of darkness, turmoil, fear and hate. I needed outside help to deal with the situation and at the time it was explained to me by my energy healer that he had been open to certain spiritual energies which had attached to me as I was in a light place. In this place I fought daemons, nightmares and negative thoughts the like I had never encountered. It was as though I was experiencing someone else's dark pain and inner hell. I meditated with pure light every day to counteract it and was completely knocked off my emotional axis. It took a lot of work and time to regain my equilibrium.

This area can be explained as being in touch with the dark passionate, animal side of our psyche. That place where lust, power, greed, primal instinct, destruction, hate, violence and passion are all coming from. Where there is no guilt, no shame or remorse and power comes from being in control and doing and taking from life that what one requires.

The in between place is where we as humans were created to live and play. When we understand that our depth and breadth of emotion is that which makes us human, that which makes us divine and that which makes us creators with the potential to do or be whatever or whoever we want. In this place we find most of humanity at different points and degrees. We are the cosmic balancing tool, the litmus paper in creation and the direction that the future of life will be taking. Our purpose therefore, is to maintain and control our lives in a way that sets a template for how we see the progression of humanity. Because it is in doing our work on the physical spiritual and mental that we raise the standard of our existence.

What is the role of happiness?

If you consider our place in the universe as like being on a large XY graph. On one side of the x axis we have the light pure mind non emotional logical spectrum and at the other end of the x axis we have the dark primal and animalistic end of existence then we are somewhere near the middle toward the vertical y axis.

Along the vertical y axis we have emotions and feelings which produce an arc or spectrum of frequencies through which we experience our reality. Emotions indicate how we experience reality. Our depths of feelings combined with our power of mind are the elements with which we create and manifest existence. Therefore in understanding this we become more effective in choosing how we show up and create our life.

Happiness, an emotion, a choice, a human condition where we experience higher vibrational feelings, a love of life and of those around us and very often appreciate and experience life in a very positive and intense way through people, places, emotions and absence of negativity because our reality and emotional being becomes mirrored back to us.

Generally a place where we feel good.

This place of happiness is an emotion. Psychologists and philosophers have documented that we can both decide to be happy which is directed from an internal philosophy and decision or we are reactive to happiness thereby placing our emotional wellbeing onto another therefore making that person responsible for our emotions.

When we create with our emotions and thoughts we will bring in that which is of a vibrational match to the frequency we send out. Therefore if we operate at a lower level of frequency and emotion we will manifest a matching reality. And same goes with higher emotional states.

2. What about Love?

All about Love

Love starts with ourselves.

To feel good we need to love ourselves, body – mind spirit – divine triangle of wellbeing.

Treating our body with love from a cellular level up. This starts with what we put into our bodies to how we exercise, discipline ourselves, groom ourselves, look after our environment, who we have around us in our life, how we allow others to treat us and how we treat ourselves and including allowing and accepting help from those around us when we need it.

When we neglect our own body and life we fail to attract into our lives that which is best for our own wellbeing and this includes setting and maintaining personal boundaries.

When we operate from a place of unshakable and solid personal love for ourselves we will see that reality through happiness, joy and an enriched reality manifest through the emotional wellbeing we have created. This includes our relationships in romance, family and finances.

The love you show for yourself will reflect in the world around you from the healthy relationships, environment, and reality we experience.

The vibration we want to be operating at is from Love and above set around 500htz and is our greatest gift to ourselves and others.

3. How do I make more money/ have a better life?

   This is simple, treat yourself with love and respect, nurture yourself and your life, do what makes you happy, raise your frequency and this will happen automatically. Take action on your dreams, ideas and when you know what you want go for it without hesitation. Ask for what you want, appreciate, receive and enjoy it.

4. Finance V Service

Finances are just energy that you give out and receive. Working on loving yourself and improving your life is your service. The more you nurture and love yourself and your life the more you improve your vibration and personal energy and in turn the world around you. This is our purpose; this is why we are here, to improve our own version of the world and to learn, prosper, grow and flourish.

5. Is there a heaven?

   If you believe in heaven then it is a place
   you create via your beliefs, your emotions,
   and your thoughts while alive. It's what you
   expect it to be. As we are of divine origin
   our bodies of carbon may cease to function
   but our spirit carries on as part of the all
   one Love.

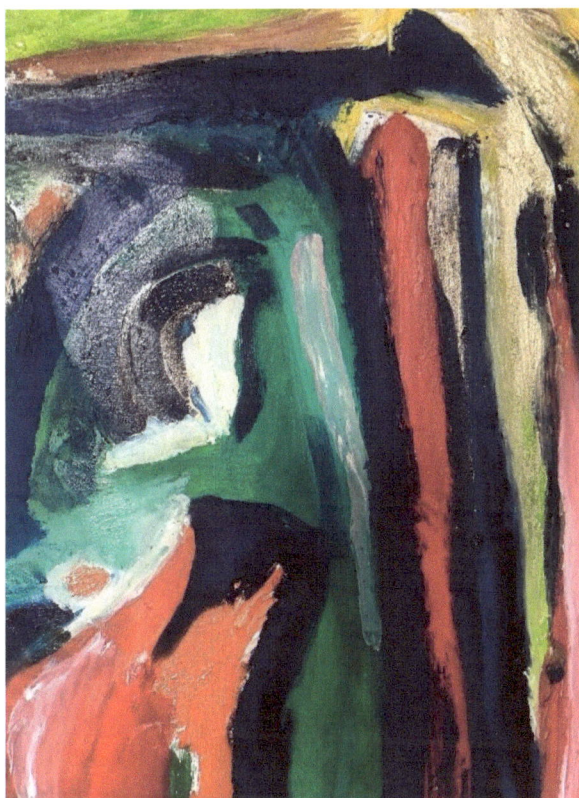

www.ingramcontent.com/pod-product-compliance
Lightning Source LLC
Chambersburg PA
CBHW041721200326

41521CB00004B/169